Edward Sayer

Observations on Doctor Price's Revolution Sermon

Edward Sayer

Observations on Doctor Price's Revolution Sermon

ISBN/EAN: 9783337098308

Printed in Europe, USA, Canada, Australia, Japan

Cover: Foto ©berggeist007 / pixelio.de

More available books at **www.hansebooks.com**

OBSERVATIONS

ON

DOCTOR PRICE's

REVOLUTION SERMON.

LONDON:

PRINTED FOR JOHN STOCKDALE, OPPOSITE BURLINGTON-
HOUSE, PICCADILLY.

MDCCXC.

[Price One Shilling and Six-pence.]

OBSERVATIONS.

Virtue itself turns vice being misapplied,
And vice sometime by action's dignified.
Within the infant rind of this small flower
Poison has residence and medicine power.
For this, being smelt, with that sense chears each part;
Being tasted, stays all senses with the heart.
Two such opposed foes emcamp them still
In men, as well as herbs, grace and rude will.
And where the worser is predominant,
Full soon the canker death eats up that plant.

SHAKSPEARE.

THE end and object of all government is the happiness of society. The happiness here meant consists not in the rude and boisterous direction of our own actions, but in the security of our natural and civil rights; namely, our lives,

A 2   in

in which is included our perfonal liberty
and fafety, our property and our good
name. This is ufually intended when a
ftate is faid to be governed by equal laws ;
that is, by laws which equally fecure thefe
three invaluable rights to all its citizens,
and which, though they reftrain our natural liberty of action in fome refpects, form
the fafeguard of what remains. The fa-
crifice is fmall, and in the fervice of the
law, is perfect freedom.

The checks impofed upon perfons intrufted with the government, by cuftom, public opinion, or by a fenate con-
fifting either of deputies, or of the whole
fociety, and in any other way that may
have been adopted by different communi-
ties, are the means by which this object of
civil government is fecured.

Thefe checks however produced, are
therefore fecondary to the object for which
they are created, and muft in themfelves,
according to their greater or leffer force,
limit the power of government in order to
fecure its proper conduct. All government is an imitation of the divine will, as

far

far as is confiftent with the weaknefs of
human nature: and God we know is a
Being infinitely wife, good, and powerful.
The governors of the earth fhould therefore poffefs as much liberty to exercife thefe
three great and beneficent qualities, as is
confiftent with the fafety of the ftate againft
the probable abufe of it.

It is obvious therefore, that a monarchy
where the monarch really acts with wifdom, virtue and power, is infinitely the
beft government for the welfare of mankind; but as it too often happens that
princes are weak and wicked like other
men, moft nations have introduced fuch
checks and limitations upon the authority
of their princes, as fuit the conveniencies
and neceffities of their refpective countries.
And in thefe checks, or the laws providing
for them, is comprized what is ufually
called the Conftitution of a State.

The beft form of government is that
which is reftrained and regulated by the re-
prefentatives of the people. At leaft the
experience of the world hitherto has given
us the affurance to form this conclufion;

nor

nor is reafon wanting to confirm it. Public opinion is too weak; cuftom, without
other affiftance, is ftill weaker. An arifto-
cracy is too ftrong, and a democracy too
wild to effect this purpofe. While repre-
fentation mixes power with wifdom, and
while it fecures and reftrains, does not too
much weaken the hands of government;
but let us always recollect, that reprefen-
tation, and the checks that flow from it,
are no pofitive good in themfelves, any far-
ther than that they are wife and proper
means of procuring it. The queftion is,
not whether they are perfect or imperfect
in their own nature, but whether they produce the end required of them, in fecuring
and protecting the happinefs of mankind
by equal laws.

If therefore under any form of government that end be obtained, the means of
fecuring its continuance are only a fecondary, though a very neceffary confideration; for while it is extremely defirable to
render the happinefs of a people perma-
nent, ftill perfons who are foberly and truly
wife, recollecting always the imperfection

2   of

of human nature, will ſeriouſly conſider,
whether preſent enjoyment may not be ſa-
crificed to an endeavour after its proſpec-
tive continuance ; and whether it would
not be better to leave future ſecurity to the
wiſdom of futurity *.

In

* To expect perfection in a government, is to expect what never has happened or ever will, becauſe it
is not to be found among the materials from which go-
vernments are formed. There is no inſtance in hiſtory
of a nation at once receiving new laws, or a new conſtitution, at the hands of a legiſlature formed for that
purpoſe ; but ariſing out of the manners, ſituation, and
neceſſities of the people, they partake of the imperfec-
tions incident to ſuch ſources of legiſlation. From the
firſt, if it were poſſible, we might expect philoſophical
and theoretical perfection ; from the latter, we in fact
receive that mixture of ſtrength and weakneſs, in which
as the one or the other is predominant, conſiſts the hap-
pineſs or miſery of mankind. The example of different nations will beſt illuſtrate a poſition that appeals to
the experience of the world. In this country, an iſland
fortified by the ſea againſt foreign diſturbance, and ſe-
cured by the ſpirit of induſtry, and by its compact ſize
from private diſorder, all our fears, and conſequently
all our laws, are principally directed againſt the ſovereign power, which if unreſtrained, might introduce
the miſchiefs of abſolute dominion. They are therefore
ſtrong and ſufficient on the ſubject of public liberty,

In this country, the end of government, which confifts in the happinefs of its fub-jects, is moft indifputably produced, and that, of itfelf, argues the fufficiency of the means which produce it. For more than an hundred years, we have enjoyed a ftate of profperity unknown to other ages and nations. The inhabitants are protect-ed in their perfons, their property, and their reputation; riches have poured in upon us from every quarter; plenty fmiles

on

while little care is taken for reprefling private enormi-ties, and lefs for repelling foreign hoftility. In France (I fpeak now of its happier hour) their fituation on the continent, and the fize of their country expofing them to great external danger, makes an abfolute power neceffary for their fafety, and the law, while it admits a great military force, and raifes an abfolute power to keep it under proper controul, forefees that from fuch circumftances private diforders would be perpetually recurring, unlefs prevented by the ftricteft regulations. In France therefore you will find little law, but fome cuftom to overawe the prince. A fpirit of military armament breathes throughout its provinces, and the police is celebrated for its competency and fuccefs. In the eaftern countries of Afia, private diforders are common, confufion of legal provifions univerfal. No regular government. The whole power is therefore

cen-

on the face of the country, and in the midſt of our wealth, we have remembered our duty both to God and man. Theſe national bleſſings are chiefly attributable to our government, and that government to the freedom and liberties of the people.

The freedom or liberties of the people are principally guarded by the frame of our legiſlature, and by the acts which have diſtinguiſhed that legiſlature for many centuries. Some defects may, upon ſpeculative principles, be found in its conſtitution, but they have exiſted from the be-

centered in the prince. Wars are frequent, and deſtruction follows them. In ſuch a ſituation the ſovereign is more than abſolute. He is deſpotic, becauſe he is the only power that can protect them againſt ſuch terrible calamities. He is for the benefit of the people deſpotic, inaſmuch as the rigour of a tyrant is tender mercy, when compared to the cruelty of a conqueror and a ſtranger. Theſe facts are mentioned not as inſenſible to the misfortunes that attend them, but as a proof of the manner in which the laws of a country adapt themſelves to its wants, and that great evils ariſing from an abuſe of government, are univerſally ſubmitted to, in order to avoid the ſtill greater of no government at all, or one not ſuited to our condition.

ginning

ginning with it. They have grown with its growth, have accompanied its fuccefs, and afford at this time no other fymptoms of decay, than they have for ages long fince paft without misfortune.

It muft likewife be admitted, that although the end of government be indeed produced by the means at prefent ufed, yet wifdom will always carry its view forward to futurity; and not content with felfifh enjoyment, endeavour, by fecuring a continuance of thefe means, to entail the like enjoyment on pofterity. But in order to make any act upon this principle neceffary, it muft be proved, that the means now ufed are deficient to the degree of doubting their continuance, which will not be eafily done, when thofe defects, if defects they are, have exifted for fome centuries, and have not failed: It muft be proved likewife, that the hazard of appealing from fettled laws to wild opinion, is balanced by the profpect of advantage, and the abfence of danger; in other words, that we do not give up a prefent benefit for the chance of continuing that benefit,

when

when it runs no great degree of rifk, and may well be left to the wifdom of the times in which it fhall begin to fail. This argument againft innovation is greatly ftrengthened by the confideration, that it cannot be afcertained, except by weak conjecture, whether thefe very defects do not contribute to the exiftence of the ftate, in which they appear, not as ulcerous blotches, but as gentle humours that pervade the whole of the body, and flow through the veins of the healthieft conftitution.

This being the fituation of our country, what motive is it that can animate the Doctor to blow his trumpet in Sion, to beat his drum ecclefiaftic, and to proclaim reforms for evils that do not exift?

Is it that the congregations of the conventicle decline in this age of peace and good fenfe? or is the money of bankrupt France like unto the money of bankrupt America, deftined only for the wages of proftitution? When no good motive can be found, it is not calumny to feek after a bad one. But to retract fufpicions of

this

[12]

this odious nature, is it the madness of his pulpit that has for ages, without a motive, sent forth its anathemas against the political peace of society?—that madness, which even Milton himself has branded as a quality belonging only to the fallen angels, a contempt of dignities and a hatred of authority.

Perhaps too it may be the Doctor's boast, as it is the boast of some men, not that they lead mankind, as Hugh Peters formerly did, through crimbledum and crambledum to heaven, and thus by confounding, enslave the understanding, but by appearing to enlarge, bewilder its progress, and perplex its operations by instruments too unwieldy for men to manage: so while one is undermining the foundation of government, and blowing up the content of social subordination, another, at the very altar of God, is handing round to his ghastly congregation the poisoned chalice of infidelity, fraught with bitter ingredients of fear, distrust, and dismay. Happy and congenial pair! the former subscribes to the government of the state only to subvert it,

and

and makes allegiance the fervant of rebellion; while the latter retains the name of
Chriſtianity the better to abuſe it, and
profeſſes religion to make the blow of infidelity more effectual.

What the motive may be, muſt be left
to Him, who is the ſearcher of hearts, and
from whom no ſecrets are hid. Its object
is not ſo cloſely concealed from human
ſight; and when the reader ſurveys the
ſplendor of the title-page; when he ſees
that the Sermon was delivered by a Fellow
of the AMERICAN PHILOSOPHICAL So-
cieties at PHILADELPHIA and BOSTON;
when he looks further and finds that this
evangelical ſcrap of political divinity is
bound up with a ſupplemental code of
French legiſlation, interſperſed with notes
taken from parliamentary regiſters, and
exaggerated accounts of *French* popula-
tion, he will be at no loſs to gueſs whence
it cometh, and whither it goeth; its *where-
about* will be inſtantly diſcovered.

It is now time to attend more cloſely to
the Sermon itſelf. We ſhall therefore
begin at that part of it which ſeems to
have

have afforded to the Doctor the higheft fatisfaction.

* He has enjoyed one Revolution, feen another, and expects a third more glorious ftill. To the two laft, as he contributed at leaft his endeavour, and perhaps fome affiftance, it is not wonderful that he fhould give the preference, fince gratitude can keep no pace with zeal heated by action. We however may form a better and milder judgment, in comparing the confequences derived to Great-Britain from what is called the Revolution, with thofe likely to be reaped by France and America from their bungling imitations. Of the latter, it would be a want of common compaffion to fpeak with afperity—Their misfortunes are their punifhment, both thofe which they feel at prefent, and thofe which are rapidly advancing. But as civil fubordination was not totally thrown afide; as the hands of government were only changed, not cut off; and as the power of the mefne lords furvived their difaffection to the fovereign; it is not to be expected that the

* Sermon, page 49.

mis-

misfortunes of that country fhould be fo marked as thofe which now form the beginning of forrows in France. Though fteeped in poverty up to the very hips; though pointed at by the flowly-moving finger of Scorn; though difcarded from the fountain of their exiftence, kept as a ciftern for foul toads to knot and gender in; yet have they not robbed the altars of the facred veffels of religion, abandoned the minifters of the gofpel to houfelefs poverty, or left the adminiftration of juftice, that great guard of fociety, to the wifdom and temperance of the mob.

To fpeak of the Revolution in France, is to fpeak of a Revolution more operative upon the manners and fentiments of all mankind, than on the government of one particular people. A whole army invited and bribed into treachery and defertion; all ranks of people levelled; all landmarks of provinces removed; property made the mock of drunken fenators; the courfe of law ftopped, and religion as it were abolifhed; (for to deprive its minifters of refpect and of the means of exiftence, is to do all that

man

man can do towards abolifhing it;)—what is this but to give a true, though a faint, picture of ferocious nature,

When wild in woods the noble favage ran;

with this great and fingular aggravation, that to the favagenefs of uncivilized barbarifm are added the vicious refinements of a fociety long enervated in the lap of peace, long ufed to lettered eafe and happinefs undifturbed. That fuch a Revolution fhould be thus fuddenly introduced, muft be attributed to the influence of very powerful and malignant principles, and be feared as the harbinger of worfe.

How different from fuch conduct was the management of that event in our hiftory, which *propter honoris caufa* has ufually been ftiled the Revolution.

The country had long groaned under actual oppreffion. The laws were overturned; the religion of the people was profaned and attacked; the property of the church was feized; the fondeft principles of the nation were infulted. After

long

long suffering and patient forbearance the whole people rose as one man. To do what?—to make fanciful improvements in the theory of their government?—to overturn every settled notion, natural, civil, and religious?—to level all ranks, and by these means bring about an universal chaos in society?—Far be such imputations from such an event. Grievances were redressed, violations of the law rectified, and the only cause of the whole was the only one who suffered from it; and even that suffering was more inflicted by his own fears, than the revenge of his injured people. The means employed to produce this effect were equally moderate with the end obtained. Subordination was strictly preserved; the reins of government never once relaxed; no treachery was used, and if it had been attempted would not have succeeded. The first men in rank and experience were obeyed by the rest from inclination. No visionary speculations (although Doctor Price's lived and wrote then to the same tune) were indulged. Past grievances were redressed as the best means

B   of

of future improvement; and whatever of
improvement followed befide, was produced by time, and followed from caufes,
which the experienced ftatefmen who laid
them knew would be certain, and only
certain as gradual alterations brought
about by events, not arbitrarily impofed
by difcordant fpeculation *. The people
were equally good and wife with their
leaders. No hatred of fuperiority, no difaffection to the higher orders of fociety,
and no luft of revenge were difplayed.
All attempts to introduce confufion were
prevented. The army, indeed, that King
James had collected together was injudi-
cioufly

* The extreme caution of our anceftors in not giv-
ing way to a fpeculation of improvement, has been
mentioned already. It is obfervable, that in the Bill
of Rights, which is confidered as the key-ftone of our
conftitution, no new right or liberty was eftablifhed
which had not before been enjoyed by the people, both
de facto as well as *de jure:* and almoft all the improvements that have followed from the Revolution, have
been made many years after it, and have followed rather as the refult of convenience than right. Thus
the dividing the expenditure of the ftate at the Revolution into the civil lift and the public expence, though
it

cioufly difbanded by their general, Lord
Feverfham. The report caufed an univerfal alarm, and the leaders of the nation punifhed the hafty indifcretion of the
general by confinement. Thus what in
France has been encouraged was re-
probated here, as the certain way to
overturn government, in feeking to re-
form it. No rebellious relaxation of
principle in the lower orders of people
was encouraged or exifted. Notwithftanding the actual oppreffion of James,
and what is worfe, the abufe of their
fondeft prejudices from ftronger prejudice
of his own; notwithftanding the difcou-

it feemed to point merely at a regulation of revenue,
has had the effect of obliging the King to hold a Seffion
of Parliament every year; which obligation, had it
been propofed as an abftract queftion, might have met
with much difficulty. So little did the leaders of the
glorious Revolution indulge themfelves in probable
improvement, that the prefs was even after that period continued under the oppreffion of a licencer, by a
new law; and on the only topic which called for ab-
ftracted difquifition, their debates have been cenfured
as exhibiting rather the jargon of fchoolmen, than the
reafon of politicians.

B 2   ragement

ragement fhewn to Proteftants, and the eftablifhment of a public mafs-houfe in the camp, with other grievances that ftrike the fenfes of a nation ; yet when the Prince of Orange had confiderably advanced in his expedition, and was enabled to hold out the protection of a well-difciplined army, no Englifhman deferted their colours, their mafter, or their oath. Lord Cornbury, a man of high family and great worth, attempted to carry over his regiment to the Prince. Knowing their hatred to defertion, he was obliged to debauch them by ftratagem into the enemy's quarters. It was then as dangerous to return as to go on. Out of the whole party thus betrayed and furprized, many of the officers, indeed, but very few of the common men, imitated the example of their commander. In this inftance appears the true morality of the Revolution. Officers had been equally fworn to unconditional obedience as well as the common men ; yet as fuch a tie is fubordinate to others of higher confequence, they did well to break through it, when expofed to that alternative,

4

alternative, of whofe delicate nature their improved educations and ftate of life enabled them to judge. The common men have indeed the fame duty, but not the fame knowledge. If they break their oaths to one mafter, they will to another, and be faithful to none. Hence if they are virtuous, they are flow to act at all; and when they do, for the moft part follow the example of their officers, whofe fortunes they fhare, whofe judgment they can rely on, and have been ufed to obey.

How differently do they manage thefe things in France!—They invert the order of events, and the evils that caufed our Revolution are themfelves caufed by their's. The object in that country is, without injury or oppreffion (other than proceeded from the nature of their government) to make a total revolution in their laws and manners, while the means they employ for that end are anarchy, or a diffolution of all civil ties, by putting arms into the hands and metaphyfical ideas into the heads of the people, a great and very great majority of whom can neither wield the one with

B 3   prudence,

prudence, nor apprehend the other without madnefs.

While we thus freely fpeak of what is
paffing among the French, as they ludicroufly ftile their nation, let it not be
thought that we approve of the abufes
that were faid to exift in their monarchy,
Abufes there certainly were, and there as
certainly were means of reforming them,
without bringing down the whole frame
of government upon their heads. As far
as fuch an object was really intended by
them, it was highly rational and defirable.
The evils which have been ufhered in under that difguife, are thofe only that we
have exclaimed againft. The blow that
affaulted the Baftile would have been ap-
plauded, had it not at the fame time levelled all the bulwarks of fecurity and peace.

A revolution thus highly feafoned cannot but be gratifying to the heated appetite of the Patriot Divine, not only as it
removes the fences of authority, and over-
turns the hierarchy of the ftate, but as it
opens into life thofe feeds of rank abufe,
which have long been planted in the bale-

5   ful

ful garden of falſe philoſophy; feeds, that left to thrive and flouriſh in their native air, give ſhade and nouriſhment to the world; but come out from the laboratory of the philoſopher—the bittereſt poiſon.

The great principle of this falſe philo-ſophy conſiſts in ſuppoſing, that the higheſt refinement of abſtracted truth is fit to be applied by all mankind to the offices of common life. It appears in almoſt every page of the Sermon under different co-lours, and with varied forms of expreſſion; but upon the whole we may pronounce it to be virtue puſhed to an extreme, from whence reſults an evil different from, though equally great with ignorance, its oppoſite? Superſtition and incredulity are indeed extremes; the one is produced by too much, the other by too little know-ledge. The firſt is the vice of a barbarous age, the latter of declining refinement. Hitherto we have ſuffered much from the former, and now (ſome ſmall interval be-twixt) we ſeem to be approaching faſt to the other. Maxims that were formerly confined to the ſchools and cloſets of the

learned,

learned, are now handed about the world as containing principles of civil conduct for all mankind; maxims, which it is impossible they should understand, unless they were all truly wise, or practise without being perfect.

It is hard to combat positions that come under such an assuming shape and appearance. There is little doubt, indeed, but that these maxims, clearing away human imperfection, are fairer in speculation than any that are now in use, as applied to common life; but then it must be remembered, that they are not fitted for action, that they are too finely spun for common eyes, and that they are therefore incompatible with the practice of mankind; though, as questions of science, they might well be discussed by such persons as could manage them with skill, since from discussions of that kind something useful may be struck out, even from materials that would be pernicious themselves in other hands. The telescope opens the wonders of new worlds to the eye of the astronomer, and while it gratifies curiosity, enlarges science, and enables him to confer fresh benefits on those,

who

who, ignorant of the procefs, enjoy the
effect. But is it therefore to be wifhed,
that Providence had invefted the naked eye
with the fame elongation of fight, or
would it not have made us ftumble over
ftraws, under our feet, while with diftant
and elevated views we were furveying the
altitude of mountains? By attempting to
remove the neceffary ignorance of mankind, that generally difplays itfelf in reve-
rential wonder, we remove the fhade from
the light, and blaft our eyes with its ex-
cefs; we ftrip the blooming fruit of its
infect blufh, as ufelefs and pernicious; like
a bafe empyric, we rob the ftone of its
flimy coat, till the tortured patient dies
under the decreafe of his malady; and to
fum up all, in clearing the eye of the mifty
film that guards it, we deftroy the fight
itfelf.

This fpecious defign of making mankind
act upon principles fo refined, that they
muft miflead, but cannot govern them,
even if they are underftood, is fraught
with every poffible mifchief; and being the
only oftenfible caufe of the late revolution

in

in France, fills the mind with alarming
apprehenfions for the peace of the world,
left the poifon hitherto confined to the
diffentions of gownfmen, fhould have diffufed itfelf amidft the mafs of mankind.

People who fpeculate, feeling the neceffity of change to which the frail nature
of mortality is conftantly expofed, have
been led to conjecture by what means this
age was to be fucceeded by another, and
how this enlightened ftate was to be followed by its companion of darknefs, fince
neither Goth nor Vandal remained to execute it.   In the moral as in the material
world, there is a conftant fucceffion of day
and night.   All hiftory proves it.   One
day telleth another, one night certifieth
another.   As Rome rofe on the ruins of
Greece, fo modern Europe fprang out of
fallen Rome, and muft itfelf fall and give
way to others.   The event feems to be certain, though the means may perhaps be
difputed.

What, however, neither Goth nor Vandal remain to do for us, we may do for
ourfelves. Such convulfions as thefe in
France,

France, aided by fuch writings as thofe
we are now canvaffing, will in the end fap
the foundation of government, by deftroy-
ing the influence of opinion upon which
it principally depends. For it muft be
fomething more than actual force, that
makes the many fubmit to the few, even
for their own advantage; and in viewing
this firft open avowal of levelling principles
in France, I cannot help fancying myfelf
in ancient Rome, reading the account of
the firft northern irruption, and predict-
ing, that what was formerly flight fkir-
mifhes on the frontiers, eafily repelled and
confined to that diftance, but which had
then become a powerful invafion, would
in a courfe of time, and after frequent re-
petitions, bring about that fall of empire
from which not even empires are exempted.

The prefs, that has yielded fo many bleffings to mankind, is the chief inftrument
employed for the propagation of thefe new
principles, and by inftilling maxims of
this kind into minds not prepared to receive them, much the fame effect is produced,

duced, as at the firft tranflation of the
bible. Though nothing but truth was
opened to the world, yet it was above the
common underftanding; and left to their
own fuggeftions, the people of thofe
days, affumed with the names of the Jews
in the Old Teftament, their character, paffions and power. Hence cities were ftorm-
ed by divine command, and men, women
and children, were maffacred agreeably to
what was then thought, the Word of God.

As the prefs is the medium through
which the poifon is diffufed, fo are declamations againft old cuftoms, prejudices
and errors, fome of the ingredients that
help to compofe it; declamations, that
are fpecious and fenfible, and would be
convincing, were men lefs men than they
are. Thus the lively fenfe of honor that
defcended to us from our warlike anceftors,
that high, noble and animating quality,
which foftens the hardfhips of life, corrects
the vices and exalts the virtues of fociety,
and fupplies even the deficiencies of religion. This proud fpirit of ancient chi-
valry and modern honor, has fooneft felt
the

the blows of unfeeling scepticifm; that
spirit which even Philofophy herfelf has
dreamt of, and has prefixed to a code of
morality *; thefe venerable checks of
confcience are now nearly removed, and
exift only in recollection.

All our old prejudices, harmlefs and neceffary, exift no longer. Thofe foft delu-
fions, that cheared us in our afflictions,
chear us no more. The holly and the
mifletoe no longer garnifh the hall of
laughter and plenty; vain pageantry!
even kiffes under it are kiffed no more;
promoters of vice, and unfubftantial plea-
fures, that hang not on the lip! The
fame dry and unfeeling ideas are extended
to other innocent frivolities, that are more
interefting to the peace of mankind; and
the mind is left bare to the cold impref-
fions that reafon may make upon it, while
venerable errors are overturned, as the cathedrals of ancient days were overturned
by the ravages of rebellion, and were ex-
changed for the *reafonable* convenience of
a modern chapel. Cathedrals, that by

* Paley's Moral and Political Philofophy.

fhed-

fhedding a dim religious light, and length-
ening out a folemn ftillnefs through their
arched vaults, make even ftone and mortar
fubfervient to the purpofes of morality.

However unpleafant and unprofitable
thefe leffer filings of modern philofophy
may appear, yet as the great attempt made
by its fupporters is directed to the fubver-
fion of civil government, fo fhall we imi-
tate them, and the Doctor, in confidering
it, as principally applied to that fubject.

We have feen how fatal all attempts at
weakening the hands of government muft
prove to the happinefs of mankind, and
may conclude from thence, that they are
forbidden even by the earlieft law, the law
of nature. Nor is revealed religion filent,
however weak its voice may vibrate in the
ears of modern philofophy. Every page
of the New Teftament inculcates the pre-
fervation of peace and fubmiffion to our
fuperiors ; not that it entirely forbids refiftance, but agreeably to the foundeft wifdom, confiders cafes that may require it
as above any defined notion of law, as
ftanding upon their own particular grounds;
and

and while it recommends obedience as generally preferable to refiftance, does not prefcribe bounds to thofe precautions which felf-prefervation neceffarily requires, and which the neceffity that creates them muft limit.

In thefe maxims is included the love of our country, which is beft fhewn as a general rule of action, by an adherence to its laws, and the fpirit of its government. The Doctor thinks proper to make an apology for the filence which he falfely fuppofes the Gofpel to maintain relative to this duty * ; namely, that if fuch a duty (though neceffary) had been inculcated at the time of the Chriftian difpenfation, it would have produced more harm than good in the then ftate of things. Oh arrogance, above all arrogance ! that it fhould be faid, and faid by a prieft in his pulpit, that the Almighty refrained from inculcating a duty neceffary to mankind, becaufe in the then ftate of human affairs, it might have been attended with fome political inconvenience arifing from the abufe of man,

* Sermon, page 7.

which

which abufe is in no fhape proved or af-
ferted to have been neceffary. What is this
but to make the wifdom of the Doctor
the ftandard of divine wifdom, and his
power the power of the Almighty. But while
thefe impious words fell from his tongue;
(and I truft they efcaped rather from omif-
fion than defign) he fhould have remembered the example he praifed, and have
omitted to inculcate even neceffary doctrine, if capable of introducing mifchief to
the world; recollecting, in the words of
his text, which he foon rambles from;
that when peace is within the walls, and
plenteoufnefs within the palaces of Jerufa-
lem, the end of government is anfwered;
and the duty we owe to our brethren and
companions, fhould, under fuch circumftances, make us add, peace be unto thee!

Having thus viewed the Doctor glory-
ing in the revolution his falfe principles
have contributed to promote, and in the
progrefs of thefe falfe principles, as pro-
moted in return by that revolution, let us
next advert to him in the act of advancing them, by a new application to principles,

ples, that by violating the order of nature, repugnant equally to the voice of
reafon and to the word of God, and oppo-
fite alike to the happinefs and to the practice of mankind, can hardly be deemed
fincere even in the perfon who fupports
them ; and the lefs fo, when, prior to this
fecond confideration of national events, we
examine a few mifcellaneous obfervations,
which like fhort words dropt in hafty
converfation, open thofe fecrets of the
heart that might have been concealed in
more ftudied gloffes ; from whence it will
appear, that the tyrant is more the object
of his diflike than the tyranny ; that his
violence peeps out from under the cloak
of affumed reafon ; and that however ill
difpofed to fubmit himfelf, he can arro-
gantly affume dominion over others.

In one of his firft pages *, this minifter of the prince of peace, with great
humility, violates by a hardy affertion,
one of the firft principles of our conftitution, and makes the king refponfible to his
fubjects. If it is meant that he is in con-

* Sermon, page 23.

C   fcience

science responsible to himself, and them, it is granted; but if the Doctor means, what the words and context imply, that he is personally, civilly, and in due course of of law responsible, then the first and most important principle of our constitution, which for the best of all reasons, preserves the personal safety of the sovereign, is violated. Every lawyer's clerk is familiar with this maxim, and the assertion that contradicts it, can alone proceed from some other motive than ignorance, and for a worse purpose than instruction.

Again, the sovereign is said to be the servant of the people *; he is their servant only as their sovereign; and to make a distinction between them, is to make one without a difference; being the sovereign of all, he is the servant of none in particular; and as sovereign, he represents them all. He is therefore his own master, since servitude consists in the power one man has over another, and no person has power over him. The obscurity of these terms gives the Doctor an opportunity to

* Sermon, page 26.

mislead

miflead the underftanding, by ufing fuch
as are difgraceful in found, not fenfe, and
by thefe means to leffen the character and
dignity of his king.

The Doctor can quote Scripture, though
he difdains to preach it. " Now letteft
thou thy fervant depart in peace, fince his
eyes have feen thy falvation," is the en-
raptured exclamation of the prieft of
peace, at the profpect of national happinefs in France *. If it means any thing,
it means, that the benefit derived from the
two events thus compared is equal in his
eyes, fince caufes that produce equal ef-
fects, are as to that effect equal themfelves.
The fimile I fhould rather chufe to apply
to the event than to the perfon ; for Doctor Price poffeffes the real modefty, as well
as the true fcience of ancient Philofophy.
But where was his ufual prudence, when
the religion of the place was profaned,
and the fanctity of the pulpit defiled, by
comparing the bleffings of the Gofpel with
the peftilence of the national affembly—
falfely fo called ? To be ferious, when

* Sermon, page 49.

C 2   Our

Our Saviour appeared the Heavens were
opened ; at his awful prefence the ftars of
the firmament fhone forth a new fplendor;
and at his coming, life and immortality
were brought to light. By the other, darknefs overfpread the land, mutual diftruft
poifoned the fources of human happinefs, and
difcontent raifed its afpiring, withered, and
threatening head. To hear fuch blafphemy,
fuch abfurdity, and fuch iniquity of fpeech,
who would not have been moved ? Even
the drowfy conventicle itfelf muft have
fhook with horror, though made up of fincere canvaffing candidates, devout placemen
in reverfion, and fanatical fools, whofe fin-
gularity forms their only deftinction.

The malignant triumph increafes with
the progrefs of the fermon, as a procef-
fion fwells with increafing fplendor at its
end, and the whole foul of the preacher
burfts forth in rapture, when he fees " their
king led in triumph*." That king, whom
even Frenchmen love, or fay they do, and
whofe good intentions have as yet been
untainted by the breath of calumny.

* Sermon, page 49.

Thus

Thus does the spirit that hates a supe-
rior, gratify itself at the expence of every
tender and humane feeling, and is accompanied with a violence that seeks to infuse
a similar hatred into others. This spirit
animates the whole sermon, and is no
where to be found more conspicuously than
in a grave passage, that informs the congregation they cannot be zealous *enough*
in promoting the cause of liberty *. How
absurd and pernicious are such maxims,
when delivered from the pulpit, without
qualification or restraint. What is it but
to invite the better part of the audience
to tumult at an election or an assize, and
the lower order to break open the gaols,
and to block up the House of Commons.
Passive obedience in the last century, was
the doctrine of the universities, and was
condemned, because it laid down a general rule without qualification, of which the
exception provided for resistance in particular cases, formed a material part; but the
doctrine of the tabernacle improves upon
this abuse, and lays down the exception

* Sermon, page 20.

C 3   of

of refiftance, without ever regarding or mentioning the general rule of obedience.

The overbearing infolence of fedition that commands a refiftance of authority, is the traditional, hereditary, and indefeafible prerogative of the conventicle throne. Nor has this tyrant of an hour neglected to exert it, as will appear from confulting the pages of his fermon; the hiftory of his *humble* defpotifm. One word will be enough to the wife. He directs his congregation to believe him. Perfuafion will not perfuade them, authority will not bind them, but Doctor Price *directs* them, and they obey *.

+ Having thus made a few mifcellaneous remarks upon detached paffages, we will proceed to confider the reafoning of the Sermon, as applied to a national event in this country, and for which the reft of it feems to have been printed, namely, a confident affurance in the repeal of the Teft Act. It has been feen, that this truly *illuftrious Apoftle of Liberty*, as his friends have fomewhat ludicroufly and irreverently

* Sermon, page 35.   + Ibid. p. 34.

called

called him*, expects the third Revolution to be extended over the whole world, and particularly over the narrow ftreights of Dover and Calais. French principles, and Diffenting want of them, may do much; but nothing will, in the opinion of fome people, contribute fo effectually to the falutary purpofe of general confufion in this country as the repeal of the Teft Laws; laws long cherifhed by a nation that knows fomewhat of its own intereft, as the guard and fecurity of their civil rights, fo far as religion is connected with them.

What other purpofe can the Diffenters propofe to themfelves in promoting the repeal? Does the law cripple them? Are they excluded by it in point of fact? Are they not to be found in our army, our offices, and our dignities, nay, in our very church itfelf †? Are not our corporations filled

_____

* Duc de Rochefoucault.

† Many chapters contain a *verfe* of conforming integrity; but in one, not an hundred miles from Weftminfter, the Diffenters, fome years ago, infifted upon nominating a Member, which was fubmitted to by the

Minifter

filled with men of such active spirit? The
law, notwithstanding this abuse, answers
a good purpose to the state, by forcing
such who entertain opinions pernicious to
the commonweal to dissemble them, when
it cannot totally prevent their entrance.
At the same time it inflicts no injury on
the Dissenters, because it does not, in
point of fact, exclude them, even if that
exclusion was an injury. The honour of
trusting to indemnifying * acts, or in dissembling their faith, is for them only to
consider, and cannot be deemed by them

Minister of that day, in opposition to the strongest
claims of church propriety. He still feeds in the same
stall.

* When any doctrine is said to be pernicious, it
should be understood to mean destructive of the principles of government. Whether it is bad in itself, or
not, is a question foreign to our purpose, and perhaps
above our decision. So let me take this opportunity
of adding, when the *Dissenters* are mentioned by name
in these observations, those only are meant who make
their faith a stalking-horse of worldly ambition; not
such as, equally interesting in number and character,
worship the universal God in the way they think best.

disgrace-

difgraceful, or they would not comply with it.

In examining the arguments for a repeal, I think myfelf bound to be concife. The fubject has been exhaufted by the contefts of a century, and by the writings of hundreds. Nothing having paffed to vary its nature at the prefent period, I fhall not, for one, think myfelf capable of altering or increafing the wifdom of fucceffive ages: but if modern wifdom be defired, let us turn our eyes to the real champion of the Church and of true Religion; a Minifter honeft at leaft, if unfortunate; though fallen on evil days, yet venerable in ruins, and great in defpite of misfortune *.

All government we have feen, and muft have noticed from our cradles, depends more upon the influence of opinion, than the terror of actual force. Religion lends a weight to opinions, that cannot be produced from any other fource. The confequence is, that its holy minifters are generally intrufted with the education of

* Lord North.

youth,

youth, in order to give the fanction of
religion to the reafon of morality; upon
both which accounts the religion of a
country and the minifters of it, fhould be
fuch only as entertain fentiments congenial to the government of the ftate; in
other words, there fhould be an Eftablifhed Church.

It will not be neceffary to prove at full
length, that the principles and doctrines
of the Church of England did at all times,
and do ftill more at prefent, fupport by
their influence the true fpirit of our mixt
and limited monarchy; or that the opinions of the Diffenters incline more to a
republican form of government. Hiftory
will prove both as a fact, and that fact
will be explained upon reafonable grounds
by a review of the faith and writings of
each. If any alteration has taken place
in the fentiments of the latter, fince they
rode triumphant in fubftantial dominion,
it has been favourable to thofe latitudinary
principles that extend the wide idea of a
democracy, to the ftill wider idea of no
government at all.

The

[43]

The reasonableness of an established church is so clear, that in many countries, and likewise in this, attempts have been made to drive the whole people into an unity of religion. However desireable that end may be, yet if the means intended to produce it intrench upon the independence of opinion, they cannot be held in too much abhorrence as the worst of slavery. The end of Government is answered, as far as it can be answered with propriety, by not permitting the influence arising from its authority to be placed in hands that will, by means of such influence, give effect to opinions destructive of the very government that protects them. No injury is offered to the persons excluded. Their exclusion is their own act, and proceeds from a restraint which every government has a most indisputable right to impose. This right is acknowledged by the Dissenters themselves, who have not even proposed a repeal of the Test Law in favour of the Papists. The right is therefore acknowledged; yet it must be confessed, that the grounds upon which that

9   right

right ought to be exerted are difputed; but, as it ftrikes me, with little difference, fince the doctrines of the Church of Rome are fcarcely lefs abhorrent to the fpirit of our monarchy than the opinions of the Diffenters; and if fupremacy be neceffary, what matters it whether a foreign fupremacy be allowed, or no fupremacy at all?

It has been faid, that no Teft being required from Members of Parliament, it is abfurd to require one from thofe who execute the law, while thofe who make it are exempted *. This objection proceeds from an inattention to the principle upon which it is impofed. Members of Parliament have no office as fuch out of their own Houfe; they therefore cannot lend the aid of the executive government to the diffeminator of principles pernicious to the ftate. But it will be faid, they can do it more effectually in the Senate. It is for that very reafon they are permitted to fit there. In that affembly all religions ought to be heard, and every fubject fhould have its fupporter. Nor is it attended with

* Sermon, page 39.

any

any inconvenience. The Diffenting Members can never do mifchief in that place, or rather do any thing, unlefs they are a majority; and if they are, according to the principles of our Conftitution they would ceafe to be Diffenters; would no longer be Prefbyterian, Anabaptift, Quaker, Lutheran, Calvinift, Arminian, Socinian, Independent, Fifth Monarchy Men, &c. &c. but by becoming the majority of the Senate, the conventicle would become the eftablifhed church.

Perfons who know what weight even the appearance of official fplendor carries with it, from the Lord Chamberlain at court, to the Beadle at church, and the ftill greater weight their power adds, will eafily fee the reafon why none fhould be invefted with either the one or the other, whofe principles and fentiments differ from thofe of the ftate, by whom both of them are conferred. If it were otherwife, the gaping multitude might follow the Mayor to the meeting-houfe inftead of the cathedral; and the line once broken, the Methodift chapel, or the Quaker's tabernacle,

nacle, would equally hold the infignia of authority with the more reafonable Prefby-terian meeting. Jack would ride three horfes at a time, and eat cuftards of all ingredients *.

It is faid by the Doctor, that the emperor Jofeph has invited the Jews into his dominions, and that France is bleffed with a Proteftant Prime Minifter †. How happy the emperor Jofeph has proved in his attempts at philofophical improvement, hiftory will tell. Brave Brabanters! let not my page be ftained with ridicule upon you, WORTHY PATRIOTS, who are defending with your blood the laws and the religion of your anceftors, confequently your own peace. May your efforts be fuccefsful! but that they may be fo, difdain to follow the example of your be-

* Addifon, who was no mean judge of the fprings of human conduct, introduces the Freeholder into St. Paul's Cathedral at the time of divine fervice, who fees, to his unfpeakable fatisfaction, the mace and the corporation (of whom not above two or three were afleep) moft devoutly prefent. Vide Freeholder.

† Sermon, page 38.

wildered

wildered neighbours, and be contented with the laws your parents have enjoyed. As to Mr. Necker, if a tree is known by its fruit, we fhall not envy France the bleffings of univerfal anarchy.

It has been faid in addition to other arguments, that the Teft Laws impofe a hardfhip on the Diffenters. The right of the ftate to make them has been proved already, has never been deniod, and is admitted by the Diffenters themfelves; the propriety of its exercife in this inftance, muft depend upon the confiderations before ftated. If it be a hardfhip, they themfelves impofe it upon themfelves; and a much heavier hardfhip would it be upon a vaft majority of the nation, to have an Anabaptift Firft Lord of the Treafury, and a dry Quaker Mafter of the Revels; or to fpeak with more ferioufnefs, the very fpring and fource of their political fafety diverted from its original and eftablifhed courfe.

* Ireland and Scotland, it is faid, have no teft, nor will it be denied, but that the fituations of fome countries may be fo pe-

* Sermon, page 37.

culiar,

culiar, as not to require it, efpecially in
countries dependent upon others, whofe
religion, or mode of faith, differ from that
of the fuperior country. In both of the
nations beforementioned, the ftate of the
Eftablifhed Church is peculiar. In Ireland
the church, though not including a fifth
part of the inhabitants, is, notwithftanding, the eftablifhed church, becaufe it
agrees with that of the fuperior kingdom,
for whofe fafety, including the fafety of
the whole, an uniformity upon that head
is procured ; but under fuch circumftances
it would be a great injuftice to make that
religion, a religion excluded from the offices of the ftate, which ought upon *general* principle to have been the eftablifhed
one ; a middle way is therefore purfued
between exclufion on one fide, and eftablifhment on the other. As much benefit
is obtained from the principle of uniformity
in religious matters, as could be obtained
without injuftice. In Scotland the eftablifhed church differs from that of the fuperior country, and was formed out of the
ruins of the latter ; it is therefore but rea-

fonable

fonable, that it fhould not be excluded as well as overturned, and in point of fact, the one was a price for the other. Till therefore the Church of England can be proved to ftand in a fimilarly. peculiar fituation to thofe of the two other countries, no arguments drawn from them are applicable to it *.

One argument more I am bound to notice, becaufe it involves an accufation of *iniquity* in the Church of England †.

Modeft

---

* In point of fact, no teft is required in either of thefe two countries, but this does not proceed from the reafonablenefs of the principle, but from the convenience of political events. Ireland was formerly a dependant, and is now a protected country. It is therefore but reafonable, that the eftablifhed Church fhould, by its doctrines, contribute to the fupport of the whole. Scotland was formerly an independant country, and is now, though united, protected by the more affluent country to which it is joined ; ftill it retains a church of its own, although of a modern date, becaufe from hiftorical reafons fuch an eftablifhment became neceffary, but the example of both countries contributes nothing to the general reafoning on the fubject; becaufe the effect their ecclefiaftical regulations may have upon their civil management, is

D   fcarcely

† Sermon, page 36.

Modeſt and humble word! but Doctor Price is a diſſenting *Pope*, and his libels are *infallible!* It is ſaid, that the participation of the holy ſacrament is made a qualification for Rakes and Atheiſts to attain the offices of government; and by ſuch means what was intended by Our Saviour to be the pureſt and moſt abſtracted rite of religion, is turned into an intereſted tool of corrupt ambition. Omitting to regard the decent, charitable, and truly chriſtian aſſertions thus made by the Doctor without proof, and which can only in juſtice apply upon general principles to thoſe who receive the ſacrament in a mode their conſciences diſapprove, let us ſee whether he is not as much miſtaken in his argument, as he is unſupported in his calumniating imputations. Qualification means ſomething poſitive to recommend a

---

ſcarcely viſible in their protected ſituation, and is maintained in its proper ſtate by the ſuperintending influence of the ſuperior country, whoſe principles, built upon a rock, have hitherto withſtood the aſſaults of ignorance, and the ſap of ſubterraneous and miſtaken philoſophy.

per-

perfon, but the receiving the Holy Sacrament, is intended only as a teſt. The qualification, if indeed it can in any ſhape be called by that name, is, that the perfon applying for an office ſhould be a member of the Church of England, of which faɛt, his partaking in the moſt ſolemn of its rites, is the beſt and cleareſt evidence. Aſſertions, nay oaths, may lye. More trifling compliances may be aſſumed for the purpoſe, and be forgiven ; but the taking of that which, according to all modes of faith, is their condemnation before God if untruly taken, is ſuppoſed to be evidence that cannot lye or be miſtaken. The Doɛtor I am afraid knows, that to ſome indeed it is a qualification, not an incidental teſt ; that they take it for a civil purpoſe only with their lips, are true to the eye, and falſe to the heart ; I draw this fearful concluſion, becauſe his whole argument leads me to it, and diſcloſes a facility of pardon in the auſterity of the Meeting Houſe, that is not to be found in the indulgencies of St. Peter.

If

If however it be an imputation to make the fentiments and ceremonies of religion fubfervient to the attainment of civil purpofes, fo I apprehend it is not quite without blame, to bring about religious improvements by a proftitution of civil integrity. The degree of blame may perhaps be different, but that difference is amply fupplied by the difference of truth between the one and the other. Our newfpapers have lately announced to us certain affociations that have been formed by the Diffenters in feveral counties, to fupport only fuch candidates for the next Parliament, as will pledge themfelves to countenance a repeal of the Teft Law. The fact is therefore certain, and what is the inference? What is it but to make the caufe of their conventicle a qualification for offices? To take away the judgment of mankind, and to damn thofe that are not within their own pale? Furthering the caufe of what they profefs to be religion, by the worft of all bad political manœuvres?—Be this a warning to you, ye who worfhip the God of your Fathers,

thers, and let not the caufe of that Church, which has efcaped fo many perils in fo many diftant ages, perifh in your hands!

The proof that is thus given us of the means by which fome perfons would advance the caufe of religion, and of the ufe to which it may be converted, was by no means neceffary. The fame perfons have long manifefted to the world, that religion may become a mere cloak for the unrighteous dealings of men. Is one of thefe holy *devotees* a Lawyer? he has a whole tribe of clients. Is he a Doctor? his phyfic purges his brethren: nor is traffic lefs affected by the intrigues of the chapel, than law and phyfic. Too true it is, that a coincidence of opinion in religious matters, like the fpirit of party, is converted into a bond of civil connexion, with a view to the advantage of their united exertions, and is often adopted folely for that very purpofe.

After all, when we confider the imprudence of attempting any alteration in matters that recur to firft principles, and involve fo many interefting and difcordant

con-

concerns, we shall still more be inclined to forbear doing so, when not even an appearance of necessity requires it, and when, if it is done, it will be considered more as the result of terror than conviction. No persons are in fact excluded by the Test Laws, since they are daily violated as to *that* effect, though, as we have seen, without defeating their object with respect to the state. The repeal of them must therefore proceed from some other motive, than from a sense of injury that is not felt, and for that other motive we can be at no loss to guess, when the end proposed is to be effected by storming the senate, and putting the senators under duress. The means are more desirable than the immediate object, inasmuch as they lead, not to a repeal of one, but to a confusion of all law. I will not suppose, that the grievance really complained of, and proposed to be removed, is the repugnance that may be felt by some persons in dissembling their faith for their interest, and in submitting to the disagreeable necessity of concurring with a great majority of their fellow citizens,

zens, in contradiction to their own *better*
opinion. I will not fuppofe this, becaufe
I fhould be forry to attribute fo much in-
fincerity to any one without proof, and becaufe if it fhould exift, the difficulties that
oppofe its entrance into the ftate, ought
to be increafed rather than diminifhed.—
Here let us paufe! Whether the repeal
of the Teft is wife or not in itfelf, becomes
a confideration light as air, when compared to the danger of furrendering the
will of the Legiflature to ftrong impor-
tunity, repeated efforts, and violent threats.
Even if the object were ever fo proper to
be complied with, the manner in which
it is forced upon our affent, would induce
us to reject the very confideration of it.
One compliance would only beget another.
Conceffion is the parent of demand, and
demand, like a thriftlefs child, whofe
paffions increafe with their indulgence,
after exhaufting what his parent can give,
dies curfing the kindnefs that deftroys
him.

Thus having defcribed the object of
certain people, as expreffed in the various

pages of this edifying Sermon, and the means by which they propofe to accomplifh it, there remains nothing elfe to confider, but of fome antidote to repel their progrefs, and prevent their malice; to confound their confufion, to fhut our ears againft the importation of foreign principles, and at leaft to bind the *hands* of thofe falfe philofophers, who mix reafon with madnefs, and would deftroy the peace of fociety *fecundum artem.* Some perfons, perhaps, animated by too much zeal, would recommend the utmoft feverity, *enfe refcindendum eft ne pars fincera trahetur.*—Such attempts are fruitlefs: they are worfe; for in the agitation occafioned by endeavouring to effectuate them, thefe miferable people and their writings are fnatched from the oblivion that would otherwife foon overtake them, and as they become infamous, become public; a price ever ready to be paid for fuch a compenfation. Hiftory too will inform us, that the reftlefs zeal of a fanatic would run with eagernefs into the grave of a martyr, if the flame that confumed him could communicate it-
                                                                        felf

felf to the world. The Church that they
thus attempt to overturn has ever enter-
tained better notions of repelling and pre-
venting their attacks. With the true meek-
nefs of Chriftianity, it retorts foul abufe
with foft reproof, and returns bitter taunts,
nay, the fevereft injuries, offered in their
hour of triumph, with forgivenefs and to-
leration, when that hour of infolence has
given way to a period of humiliation, its
infeparable follower. "Whofoever through
" his private judgment," fay the Articles
of our Faith, " willingly and purpofely
" doth openly break the traditions and
" ceremonies of the Church, which be not
" repugnant to the word of God, and be
" ordained and approved by common au-
" thority, ought to be rebuked openly,
" (that others may fear to do the like) as
" he that offendeth againft the common
" order of the Church, and hurteth the
" authority of the magiftrate, and wound-
" eth the confciences of the weak bre-
" thren."

This paffage contains the whole of the
reafons upon which obedience to civil au-

E   thority

thority is founded, and by its plain and and obvious wifdom at once points out our duty, and punifhes our difobedience; leaving to the miferable criminal the painful confcioufnefs of having attempted, in defiance of reafon, to introduce mifery and devaftation into the peaceful habitations of his neighbours.

Indeed, while language like that conveyed in this Sermon is confined to few, and is read with rapture only by congenial fpirits, it ought not to be entirely fupprefled, even if it were poffible, but fhould be referved as a warning voice, to tell the reft of the world that fuch men are abroad, and, like croffed houfes, are to be avoided as peftilential. This wife connivance muft not, however, be carried too far, as perhaps it would be, if we were to remain filent at the prefent period, when whole nations, lefs enlightened, indeed, than ourfelves, have fallen under attacks of the like nature; and though they exhibit a picture of diftrefs to the prudent, give proof likewife of fuccefs to the wicked. Let the confequence be equal, and as it fharpens the

3   appetite

appetite of the one, may it excite the zeal
of the other! left that, furrounded as we
are by all the bleffings of peace and of ci-
vilized life, great in arts as in arms, the
envy and admiration of the world, we do
not fall from this enviable eminence, this
ftate of unparalleled profperity, into that
gulph, which opens wide for the beft and
ftrongeft eftablifhments of frail mortality:
remembering always, in the words of the
great Poet of Nature, that, proud and fecure as we may think the happinefs of
Great-Britain; yet when royal ftate is
down, when dignities are defpifed, and
offices traduced; when the vanity of other
countries is aped in this; when the time
is come to mock at form, and when riot is
our care,

" It will become a wildernefs again,
Peopled with wolves, its old inhabitants."

F I N I S.

www.ingramcontent.com/pod-product-compliance
Lightning Source LLC
Chambersburg PA
CBHW031752090426
42739CB00008B/985